Exploiting the Moon

Patrick H. Stakem

(c) 2019

Number 30 in the Space Series

Table of Contents

Introduction..3
Author...4
   Exploring the Moon..5
Should we be spending money in Space?...10
   Google Lunar X-prize..10
   Lunabotics Mining Challenge..11
   Lunar Environment..12
Deep Space Gateway 13
LOP-G..15
NASA's Lunar Outpost...16
     Lunar elevators and mass drivers 17
     A launch and landing pad..17
Transportation and Logistics...18
   Past and Current Missions..18
     Soviet Luna missions...19
     U. S. Surveyor Missions...19
     Chinese Yutu Mission..20
     Indian Mission..20
     Israel..21
Commercial Players..21
What resources can we use, from the Moon?...................................21
   Lunarcrete..22
   Solar Cells...23
   Helium-3..23
   Lunar infrastructure...25
Lunar Manufacturing...26
   Lunar base...27
The Players..28
Space Law...29
Lunar Tourism..31
Glossary..33
References...39
Resources..43
You might also be interested in some of these..................................45

"First, I believe that this nation should commit itself to achieving the goal, before this decade is out, of landing a man on the Moon and returning him safely to the earth. No single space project in this period will be more impressive to mankind, or more important for the long-range exploration of space." John F. Kennedy

Yes Sir, Mr. President.

"...I'm on the surface; and, as I take man's last step from the surface, back home for some time to come - but we believe not too long into the future - I'd like to just [say] what I believe history will record. That America's challenge of today has forged man's destiny of tomorrow. And, as we leave the Moon at Taurus-Littrow, we leave as we came and, God willing, as we shall return, with peace and hope for all mankind. Godspeed the crew of Apollo 17"

Gene Cernan, Astronaut, 12/13/72

Introduction

Besides the technical challenges, the Apollo missions were a matter of National Prestige. In the Space Race between the United States and the Soviet Union, President Kennedy said we were going to the Moon and return safely before the end of the (1960's) decade, so we did. Ok, its been 50 years since I watched the first lunar landing on black & white tv. We have put satellites in lunar orbit, and men and rovers on the surface, but we need to go back.

The space race to the Moon was driven by the Cold War which in turn drove technology development. It was an immense kick-start for the semiconductor and integrated circuit industries world wide.

With Commercial firms involved and interested in mining the moon and asteroids, Earth will have to develop a more comprehensive "Space Law" to address who owns what and who benefits. In the past, the new frontiers, America, the Yukon, the "West" were mostly wild and ungoverned, at least at first.

Hopefully, we will think this thing through, so no corporation or Nation-state will be able to enrich themselves, at the cost of others. This involves proper interpretation of the Outer Space Treaty, signed by most nations. Of particular interest are sites on the lunar surface with resources that could be mined, and areas of total or no sunlight. Looks like we will need a legal cadre to sort out the details. There is some precedent in Antarctica, and the oceans.

What is the most valuable thing on the moon? Well, if you want something light weight that is worth bringing back to Earth, that would be Helium-3, of use in a new generation of fusion reactors that would be less costly, and much less dangerous. Probably the next thing would be water, but we would use that in-situ both for greenhouses, but also cracked down, with solar power, to its constituent hydrogen and oxygen. That's rocket fuel, good for the return trip, or for going out further. That's rocket fuel that doesn't have to be carried up from Earth by more rocket fuel. Also, the oxygen supply for lunar bases can come from local sources, until we get the Greenhouses going. There's always a need for spare oxygen.

Author

Mr. Patrick H. Stakem has been fascinated by the space program since the Vanguard launches in 1957. He received a Bachelors degree in Electrical Engineering from Carnegie-Mellon University, and Masters Degrees in Physics and Compute Science from the Johns Hopkins University. At Carnegie, he worked with a group of undergraduate students to re-assemble, modify, and operate a surplus missile guidance computer, which was later donated to the Smithsonian. He was brought up in the mainframe era, and was taught to never trust a computer you could lift.

He began his career in Aerospace with Fairchild Industries on the ATS-6 (Applications Technology Satellite-6) program, a communication satellite that developed much of the technology for the TDRSS (Tracking and Data Relay Satellite System). He followed the ATS-6 Program through its operational phase, and worked on other projects at NASA's Goddard Space Flight Center including the Hubble Space Telescope, the International Ultraviolet Explorer (IUE), the Solar Maximum Mission (SMM), some of the Landsat missions, and Shuttle. He was posted to NASA's Jet Propulsion Laboratory for Mars-Jupiter-Saturn (MJS-77), which later became the *Voyager* mission, and is still operating and returning data from outside the solar system at this writing. He initiated and lead the international Flight Linux Project for NASA's Earth Sciences Technology Office. He is the recipient of the Shuttle Program Manager's Commendation Award, and has completed 42 NASA Certification courses. He has two NASA Group Achievement Awards, and the Apollo-Soyuz Test Program Award.

Mr. Stakem has been affiliated with the Whiting School of Engineering of the Johns Hopkins University since 2007. He supported the Summer Engineering Bootcamp Projects at Goddard Space Flight Center for 2 years. These resulted in the Greenland Rover, a tracked robot measuring the depth of the ice sheet.

Exploring the Moon

We humans have been looking at the Moon for millennia. The first guy to get a really good look was Galileo in 1609, after he invented the telescope. The first probe to reach there was the Soviet's Luna-2 in 1959. If you remember July 1969, then you were glued to the black & white tv, watching the first humans landing on the moon.

The Greek Anaxagoras, in 428 BC, thought the Moon was a giant rock, reflecting the light of the Sun. Good call.

The Moon was the first extra-terrestrial body to be explored by humans. It is close enough that the communication time is about ½ second, and lunar spacecraft could be controlled from Earth. But, the lack of communication with the craft when they are behind the moon, from the Earth viewpoint, dictated at least a stored telemetry and command capability. Rovers on the face of the moon towards Earth are in continuous contact.

Early in the era of space exploration, a series of rover vehicles were sent to the Earth's moon. These were designed as precursors to a manned visit. From the mid-1960's through 1976, there were some 65 unmanned landings on the moon. There was also a private effort, the Google X-prize. The moon is still the subject of intense study, with missions from the United States, Russia, China, India, the European Union, Japan, and Israel.

The Soviet Union launched a series of successful lunar landers, sample return missions, and lunar rovers. The Lunokhod missions, from 1969 through 1977, put a series of remotely controlled vehicles on the lunar surface.

The NASA Surveyor missions of 1966-68 landed seven spacecraft on the surface of the moon, as preparation for the Apollo manned missions. Five of these were soft landings, as intended. All of these were fixed instrument platforms.

Let's look back some 50 years, and review the manned exploration of the Moon. We had bigger rockets back then, but are catching up quickly, in both the government and now the private sectors.

On top of the Apollo rocket stack was the 3-person Apollo payload. The Lunar Excursion Module (LEM) was stored behind the service module. Once in Earth orbit, the capsule and Service Module were separated, the capsule rotated 180 degrees, and docked to the Lunar module. The lunar package was then separated from the third stage. The capsule, lander, and service module left Earth orbit heading for the moon, while the third stage was commanded into a solar orbit, to get it out of the way.

The Command Module, or Apollo capsule, was the cockpit and living quarters for the three astronauts. The computing heart of the capsule was the unique Apollo Guidance Computer. The need for a computer onboard the Apollo was required by the chosen approach to the mission, lunar-orbit rendezvous. Part of the spacecraft (Command Module) would remain in lunar orbit, while a detachable part (LEM) would descend to the surface. Later, the LEM would return to lunar orbit and rendezvous with the Command Module, which would then leave lunar orbit and return to Earth. The ability of the Command Module and LEM to do in-flight computations was crucial to this approach. At the time, the only guidance computers were developed for ballistic missiles.

The iconic guidance computer got the jobs of getting to the moon, landing, taking off, and getting back to Earth done. But your phone is many thousands time faster and more capable. Electronics advances by Moore's law, with a doubling of capability every couple of years.

The Service Module was located behind the Command Module, and the astronauts had no direct access to it. It was unpressurized, and contained a restart-able liquid rocket engine and associated propellant, fuel cells, and electronics to support the mission. The fuel cells used hydrogen and oxygen, and some of the oxygen was also used to replenish the Command Module atmosphere. It had a reaction control system to adjust the spacecraft attitude. The service module had radiators to dump excess heat, and a high gain antenna to communicate with Earth. The Command Module stayed attached to the Service Module until just before reentry into Earth's atmosphere, when the Service module was commanded to reenter the atmosphere independently and burn. The Service Module relied on the AGC in the Command Module for computation.

The lunar excursion module allowed a two man crew to land on the lunar surface, stay for a period of exploration, and return to the Apollo Command and Service Modules in lunar orbit. It had an Apollo Guidance Computer, programmed for the different and difficult tasks of landing on the lunar surface, and later taking off from the surface. Compared to the Launch complex at KSC with all its support infrastructure, the computer in the LEM did not have a lot to work with.

The LEM had two sections, one of which held the descent engine, and stayed behind on the Lunar Surface. The Ascent Stage, holding the two astronauts and the Guidance Computer, rendezvoused with the Command and Service module in lunar orbit. The Service Modules, returning from the moon, were jettisoned before reentry and burned in the atmosphere.

The Apollo mission Lander stages are all still there, where we left them. That's a good place for monuments. Most of the crewed portion of the LEM's were deliberately crashed into the lunar surface, to provide data for the seismic instruments left on the surface. The Lunar Modules *Antares, Falcon, and Challenger* impacted the lunar surface.

One of the later missions to study the Moon is the Lunar Reconnaissance Orbiter, LRO, from NASA/GSFC. It was launched in 2009, and is still operating. It is is a polar orbit, coming as close as 19 miles (30 km) to the lunar surface. It collected the data to construct a highly detailed 3-D map of the surface. Up to 450 Gigabits of data per day are returned to Goddard Space Flight Center.

One interesting feature on the moon is the lava tubes, long underground channels left after lava flow. They exist on the Earth as well, and Mars. These might be exploited as lunar habitats, or shelters in case of radiation storms. They also moderate the swings in surface temperature. These tunnels are located under the surface. The diameters range from 5 to 90 meters. The probably need to be lined. The lava tube gives away its position by a structure called a "skylight," a circular opening on the surface. LRO has imaged over 200 potential skylights. Gravitational measurements of the surface from orbit have shown that tubes more than a kilometer in width may exist.

Since the moon is in tidal lock with the Earth, we never see the backside from Earth. So close, yet so far away. Any rover on the lunar backside is permanently out of touch with Earth. Lunar orbiters are out of communications with Earth for somewhat less than a half orbit. This could be solved with a communications relay satellite in lunar polar orbit. The moon does wiggle a bit, less than 50% of the surface is permanently out of touch.

If we could precess the lunar communication relay spacecraft's orbit so it remains normal to the Earth-Moon axis, it would be ideal. This might involve excessive propellant expenditure, shortening the mission life.

A lunar orbiter can store data on the back side, and send it back to Earth when it's over the front side. Might be a nice job for a Cubesat. As a student project during the Summer of 2016, the author mentored an engineering team to develop a Cubesat-based rover for explanation of the lunar backside. This was a ground-based tracked vehicle. It had an associated lander/habitat where it could spend the 14-day lunar night. This prototype was tested at NIST's robotic testing lab, which usually sees urban search and rescue and bomb disposal robots.

An interesting NASA Cubesat mission, the planned *Lunar Flashlight*, involves a 6U Cubesat with a solar sail. It is scheduled to go on the Space Launch Mission, Expedition One, in 2020. It will go into lunar orbit, and look for ice deposits and potential areas of resource extraction (ie, lunar mining areas). It will be able to use it's solar sail to reflect sunlight onto the surface, particularly the polar regions. The solar sail will be 80 square meters in size.

The Orion crew module will be sent empty on its first mission, sometime in the future, that will take it beyond the moon. Orion is planned to be used for human return missions to the moon, and Mars. This first flight is called Exploration Mission one (EM-1), and is unmanned. After the Orion capsule has separated from the upper stage of the launch vehicle, several Cubesat missions will be deployed. The Orion upper stage can support 11 6-U Cubesats. The Bicentennial mission will contain the yeast *S. cerevisiae*, and will have an 18-month duration. After the Cubesats are deployed it will enter a lunar fly-by trajectory and then a heliocentric orbit where its distance to the sun will be slightly closer than Earth's.

The third payload is the Lunar Flashlight. NASA's SIMPLEx mission (Small Innovative Missions for Planetary exploration) had 13 Cubesats lined up for potential 2018 launches on an Orion vehicle. This will spend three weeks in space, including 6 days orbiting the moon. Cubesat missions will include the Lunar Flashlight, to look for water ice; SkyFire, from Lockheed Martin; Lunar Ice Cube; CUSP, Cubesat for Solar Particles, and Lunar Polar Hydrogen Mapper. More will be selected closer to launch time.

Should we be spending money in Space?

Trick question? There is nowhere to spend money in space. The International Space Station doesn't have an ATM. We spend money on Earth for space related projects. Doing this gives us a huge return. Think about weather forecasting, the GPS navigation system, and Dish Network. In the same sense, there is no way of spending money on the Moon. We can spend lots of money getting there and getting back, but that is money spent on Earth. It may prove to be a great investment.

Google Lunar X-prize

This was a contest announced in 2007 for private teams to be the first to land a robotic spacecraft on the Moon, travel 500 meters, and transmit back to Earth high-definition video and images. How hard could this be? The deadline was originally the end of 2014, but it was extended to the end of 2017. A lot of good teams around the globe worked on this. The deadline was extended again to March 2018, with 5 teams still in the running. In January, the X-prize foundation announced that no team was going to meet the launch window. The contest was closed, and no one won the $30 million.

On the other hand, this project was a great learning experience, and got smart engineers to look at the problem and potential solutions. Spin-offs from the X-prize work include the company MoonEX (Moon Express), which is focusing on mining the moon for natural resources. These include the rare-Earth elements niobium, yttrium, and dysprosium. The company has also placed a small telescope, the International Lunar Observatory on the moon. They also have a Commercial Lunar Payload Services contract with NASA. The FAA has granted them the first license to deliver commercial payloads to the moon, as required by the International Outer Space Treaty, of which the U. S. is a signatory. The also have a signed agreement with the Canadian Space Agency.

Lunabotics Mining Challenge

NASA's Mining Competition was established in 2010, and is ongoing. It challenges college students to apply system engineering principles to mining scenarios, and test the hardware in the Caterpillar Mining Area. Schedule, Budget, and Design Philosophy are parameters for judging. NASA also required K-12 outreach. There are a series of awards in different project areas. Quoting from the announcement, "The Lunabotics Mining Competition is a university level competition designed to engage and retain students in Science, Technology, Engineering and Math (STEM). NASA will directly benefit from the competition by encouraging the development of innovative lunar excavation concepts from universities which may result in clever ideas and solutions that could be applied to an actual lunar excavation device or payload. The challenge is for students to design and build a remote controlled or autonomous excavator (Lunabot) that could collect and deposit a minimum of 10 kg of lunar dirt within 15 minutes. The complexities of the challenge include the abrasive characteristics of the lunar surface, the weight and size limitations of the lunabot, and the ability to control the lunabot from a remote control center. Twenty two teams from around the nation were ready to compete at the Kennedy Space Center Astronaut Hall of Fame on May 27-28. These are annual events, with teams selected each year.

"The challenge will be conducted in a head-to-head format, in which the teams will be required to perform a competition attempt using the regolith sandbox and collector provided by NASA. NASA will fill the sandbox with simulated regolith, compact it and place rocks in it. Each competition attempt will occur sequentially. Between each competition attempt, the rocks will be removed, the regolith will be returned to a compacted state and the rocks will be returned to the sandbox. Consideration of prize awards will be based on each team's performance during the official competition attempt. All excavated mass deposited in the collector during the competition attempt will be weighed after completion of the competition attempt. The teams that excavate the first, second and third most lunar regolith mass over the minimum excavation requirement within the time limit will respectively win first, second and third place prizes."

Lunar Environment

Anything on the lunar surface operates in vacuum. Not a perfect vacuum, but fairly close. This implies a few things. Lubricants evaporate and disappear. All materials out-gas to some extent. All this stuff can find its way to condense on optical surfaces, solar arrays, and radiators.

Another issue in vacuum is the cold-welding of certain metallic materials. This occurs when two pieces of material, without an oxide layer, are pressed together. This is facilitated by having very clean surfaces, and a vacuum environment. This affects moving subsystems such as solar arrays and steerable antennas. Early deployment of mechanisms is usually not a problem, but mechanisms that have to move throughout the mission, such as solar arrays and antennas, can be problematic.

The radiation environment of Earth orbit is well understood. The crew on the ISS can operate for up to a year at the ISS, before accumulating a "life-time" dose. In additional Sentinel satellites give us a few days warning of solar storms, or Coronal Mass Ejections, that would allow the crew to enter a specially designed "storm shelter" for a few days. The situation at the moon is much different. The moon has a weak magnetic field compared to Earth. The good news is, there are almost no trapped particles, like Earth's Van Allen Belts. The bad news is, the lack of an appreciable magnet field mean no protection from energetic charged particles. In cis-lunar space, there is a greater incidence of galactic cosmic rays. For Habitats at the Earth-Moon libration points, it is the same.

Away from our home planet with its convenient magnetic field and van Allen Belts, we have radiation issues with cosmic rays and the Solar proton wind, as well as transient events such as the CME. This will be addressed by locally obtained mass, from the lunar surface or asteroid, employed as bulk shielding. The shielding will also protect against space debris. Although a window may be pierced, the hole will generally be small enough to be ignored for a while, due to the large enclosed volume of air.

Facilities on the lunar surface will probably be put under a layer of regolith. Another potential accommodation is within lunar lava tubes.

Colonies are best located near resources and energy sources. The energy resource is the Sun. This approach was followed in 19th century iron manufacturing, where the iron furnaces were located near supplies of the raw material, iron ore, limestone, and coal. Villages grew up around the iron works. It is generally cheaper to ship finished product than the raw materials.

Deep Space Gateway

The Deep Space Gateway (DSG) is a NASA Project for a crewed station in cislunar space. It is intended as a jumping-off point. The Orion crewed vehicle is scheduled to be used for this effort. The Gateway would be located in a halo orbit around the Moon. By that, we mean that the spacecraft would be visible to Earth for its entire orbital path. The DSG would form an in-orbit ecosystem for missions to the lunar surface, and later to Mars. Ion thrusters are proposed for station-keeping. These use electrical power for accelerating various (usually, inert) gasses to high velocity, rather than using fuel and oxidizer. The thrust is generally low, but can be continued for long periods of time.

The Power/Propulsion Module (PPM) will have large solar arrays for power, 40 kW is baselined. The cis-lunar Habitation Module will join the PPM later in orbit, and will provide living and work space. The planned Gateway and Logistics module will join next, providing experiment space and supplies. An airlock module will be added later, to enable EVA operations. The DSG can also serve as a communications relay to and from Earth, for lunar missions. Telerobotic missions on the lunar surface can be accomplished now, but the communications delay is just on the edge of making it awkward. The communications time from the DSG to the lunar surface will be negligible. The DSG could also serve as the basis for a lunar GPS system, providing a location reference for surface rovers.

Although we have data on 1-year duration missions close to Earth, the DSG will provide more information on long duration human missions in an environment away from our home planet. All of the ISS partners, US (NASA), Russia (Roscosmos), Europe (ESA), Japan (JAXA), and Canada (CSA) are participating, and form the nucleus of further human exploration of the solar system.

The project presents daunting challenges in design, testing, delivery, logistics, and operations. The lunar vicinity provides a effective location for expeditions to other places in the solar system. And, it's been a while since we have stepped foot on the Moon. In the mean time, the discovery of water ice at the poles, from remote sensing, is most interesting for a lunar base.

The Gateway will serve as a "enabling infrastructure" for further exploration. It is planned to be placed in a near-rectilinear halo orbit (NRHO) around the Moon The Gateway will serve as the starting point and the mothership for lunar exploration. The DSG offers return to Earth in a matter of days. If the lunar water ice can be successfully mined and broken down into hydrogen and oxygen, we will have a fueling station that is not on the Earth's surface. With lunar surface exploration by autonomous and telerobotic rovers, mining for minerals can also be accomplished.

The Gateway will differ from the ISS in several ways, but will certainly incorporated the "lessons-learned" from ISS and predecessor stations. We don't have the Shuttle anymore, so logistics will involve a new generation of heavy lift vehicles. It will probably be built in lunar orbit, so some facilities for the construction crew will be required.

The DSG will result in a one-year crewed mission near the moon, to validate the concept of a flight to Mars. It is not so much the distance to Mars, as the relative orbital positions of the two planets in their solar orbits. The mechanics of the transfer orbit were worked out in 1925 by German scientist Walter Hohmann. In his 1928 book,

The Gateway can also participate in lunar in-site research utilization, using lunar ice from the surface, brought back to the Gateway to be separated into its constituent hydrogen and oxygen, and used for rocket fuel. Hosting crewed surface missions, the Gateway would operate on a similar model to Antarctic bases.

LOP-G

The Lunar Orbital Platform-Gateway (LOP-G), is a renaming and restructuring of the Deep Space Gateway Program, a joint Russian-US effort, and associated missions. This is a step beyond the International Space Station, which will be beyond its useful lifetime in a few years, and will be decommissioned, with some parts being reused, and some re-entered. This will result in a new era of human space exploration, further from Earth. Whether we refer to the emerging facility as a Gateway, a Colony, a settlement, or a habitat, we are talking of a permanently occupied facility. We can consider the habitat to be in orbit (about something), or on the surface of another body, other than Earth. These projects will differ in detail, but will all consist of self-sufficient structures somewhere other than Earth, with an associated logistics train. The Gateway would be continuously crewed, and serves as an outpost form which to explore the lunar and Martian surfaces.

NASA's Lunar Outpost

The NASA Lunar Outpost is an element of the Bush administration's (2001-2009) *Vision for Space Exploration.* It was directed by Congress that the facility would be named the *Neil A. Armstrong Lunar Outpost.* The outpost location would be at one of the lunar poles. Since then, remote observations have revealed the presence of water ice in craters at the South Pole. The South Pole remains in shadow, and sunlight does not reach the bottom of the craters. Besides the value of in-situ water supplies, and the ability to produce hydrogen and oxygen from the water via solar-powered hydrolysis, the ice may contain records of the material of the early solar system. The water is a critical sustainable resource for a crewed lunar base. The Indian Chandrayaan lunar orbiter was responsible for this discovery.

One ideal structure for lunar bases (that the author has worked on) is lava tubes. These are found on the Earth as well, and the cooled lava provides a hard, sealed surface. It just needs to be capped with airlock doors, and no further exterior construction is required.

The outpost will consist of various modules for habitation and laboratory space, an extension solar array assembly, and a garage for rovers. There will also be a communications facility for the link to Earth. The facility was designed for a crew of four with 7-day visits during deployment, and up to 180 day operational missions.

NASA is looking to private enterprise to provide logistics support for the lunar surface, and some companies are making lunar plans themselves. Bigelow Aerospace has plans to build bases on the moon, based on their inflatable modules, originally developed from NASA technology. There is one of these attached to the ISS right now.

Lunar elevators and mass drivers

Two inexpensive options, compared to rocket launches, to get lunar material to orbit are the space elevator, and the mass driver. The space elevator is a concept that would work nicely on the moon. We have the advantage of a lower gravity than Earth, no atmosphere, and it could be built with currently available materials. There is nothing nearby to interfere with its operation. The lunar elevator would span about 50,000 km. The elevator needs a solid tether on the surface, a large mass at the upper end, for a tether, and a very strong cable. A lunar elevator could be tethered to a mass at the L1 Lagrange point, between the moon and the Earth. An elevator on the back side is also feasible. Space elevators have been explored since the 1890's. We now have the technology to construct them. A handy asteroid could be used as the counterweight for the lunar elevator.

A mass driver is an electromagnetic catapult, utilizing a long, linear motor. This would work well on the lunar surface, with its reduced gravity compared to Earth, and lack of atmospheric drag. The other advantage is 15 days of sunlight, to operate the driver as well as to charge batteries. In a mass driver, the payload does not contact the launch rail, but is magnetically levitated, which requires electrical power.

A launch and landing pad

One of the first things we will need on the lunar surface is a launch pad for the return trip. This can be constructed of lunacrete, discussed later in this book. A big flat platform to land on will help mediate that large amounts of lunar dust that will get thrown up during a powered landing. It will provide a stable platform for the landing vehicle. It can be one of the first construction projects, even before habitats.

Transportation and Logistics

Transporting material and supplies to the lunar surface is easier than transporting crew. There is a transportation infrastructure for getting materials to and from the International Space Station, separate from that of handling astronauts.

The CATALYST (Lunar Cargo Transportation and Landing by Soft Touchdown) Project is a NASA initiative to develop robotic lunar landers to deliver payloads to the lunar surface. This will be done in partnership with U.S. commercial players. The project has already signed up three company's with Space Act Agreements. Letting the delivery service go commercial, NASA can focus on the science and technology.

NASA's Commercial Lunar Payload Services Program (CLPS) solicited proposals from a group of company's for delivery services to the moon. Significantly, these are all aerospace company's, not logistics companys such as Fedex and UPS. Astrobotics is partnered with logistics giant DHL. Some of thework is a spin-off of the Google Lunar X-prize efforts, which did not have a winner. Three companies have been selected, Masten Space Systems of Mojave, California, Astrobot Technology of Pittsburgh, and Moon Express, of Moffett Field, California. Moon Express holds the first commercial license from the U.S. Government to go to the Moon.

Amazon is addressing lunar delivery, as Jeff Bezos is head of Amazon as well as rocket company Blue Origin. Amazon knows a lot about getting stuff delivered, quickly and efficiently, while controlling costs. We hesitate to refer to this as *Lunar Prime*. There is also a need for delivering material back from the moon to the Earth.

Other private company's involved include OffWorld; focusing on worker robots, and *ispace* in Japan.

Past and Current Missions

This section discusses past and ongoing lunar exploration missions.

Soviet Luna missions

The Soviet Union launched a series of successful lunar landers, sample return missions, and lunar rovers. The Lunokhod missions, from 1969 through 1977, put a series of remotely controlled vehicles on the lunar surface. Lunokhod-1 was an 8-wheeled rover, operated from Earth. It was the first Rover to land on a body other than Earth. It deployed from the landing platform via a ramp. It was operational for 11 months. The follow-on Lunokhod-2 Rover could transmit live video from the surface, and had a series of soil property instruments. Its tracks were seen by the Lunar Reconnaissance Orbiter in 2010. The Lunokhod-3 rover was built but never launched. It resides at a museum. The first and second rovers remain on the moon, although the second rover was sold in 1993 at a Southby's auction. The buyer was Richard Garriott, son of Astronaut Owen Garriott. As of this writing, he has not picked up his property.

The initial purpose for the Lunokhod series was to scout sites for manned landings, and to serve as beacons. The rover could be used to move one Cosmonaut at a time on the surface as well. Lunokhod had a group of four television cameras, and mechanical mechanisms to test the lunar soil. There was also an X-ray fluorescence spectrometer, and a cosmic ray detector. The second unit conducted laser ranging experiments from Earth via a corner reflector, and measured local magnetic fields. The rover was driven by a team on Earth in teleoperation mode.

U. S. Surveyor Missions

The NASA Surveyor missions of 1966-68 landed seven spacecraft on the surface of the moon, as preparation for the Apollo manned missions. Five of these were soft landings, as intended. All of these were fixed instrument platforms. Interestingly, Apollo-12 astronauts landed near Surveyor 3, and returned with some pieces. Not just souvenirs, these were used to evaluate the long term exposure of materials on the lunar surface.

LADEE

NASA's Lunar Atmosphere and Dust Environment Explorer was launched in 2013 to study the lunar exosphere and dust. It was developed at NASA AMES, and operated at GSFC. LADEE was to study the moon's "dust atmosphere." After the mission, it was crashed onto the lunar backside, and this was witnessed by LRO. One finding involved the gas neon in the lunar "atmosphere." This is exceeded by the amounts of helium, argon, and neon. Iron and titanium were also observed to be present.

Chinese Yutu Mission

Yutu is the name of the Chinese Lunar Rover, and means Jade Rabbit. It was launched in December of 2013. It landed successfully on the moon, but became stationary after the second lunar night. It is a 300 pound vehicle with a selection of science instruments, including an infrared spectrometer, 4 mast-mounted cameras including a video camera, and an alpha particle x-ray spectrometer. The rover is equipped with an arm. It also carries a ground penetrating radar. It is designed to enter hibernation mode during the 2-week lunar night. It does post status updates to the Internet, and still serves as a stationary sensor platform.

Indian Mission

The Indian/Russian Chandrayaan-2 mission is an orbiter and lander.. The design is unique in having been selected from student proposals. The lander will be a 6-wheeled, solar-powered rover. It is due to be launched in 2019. The previous Chandrayaan-1 mission was launched in 2008, and operated for a year in lunar orbit. It also released an impactor, which landed at the lunar south pole.

Israel

The latest member of the mission to the moon club is Israel, who launched a lander named *Beresheet* on a Space-X Falcon-9 rocket as this book was being prepared. The mission was executed by a privately funded non-Profit organization called SpaceIL. It was a fairly substantial craft, at 1,300 pounds and five feet tall. This was a continuation of the group's work on the Google X-prize.

Commercial Players

In early 2019, NASA announced that three companies were selected for lunar logistics, under the Commercial Lunar Payload Services program. This would involve only unmanned missions to the lunar surface. The companies were Astrobotic Technologies, Intuitive Machines, and Orbit Beyond. The latter company is planning a first flight in 2020, with Astrobotics going the next year. The specified areas are 4 payloads to Mare Imbrium, 14 to Lacus Mortis, and 5 to Ocenus Procellarum. Astrobotics applied its experience with the Lunar X-prize competition to this venture. Orbit Beyond is using licensed technology for the lander for Indian X-prize contender Indus. In addition to NASA payloads, the companies can carry commercial packages, on a space/weight available basis.

In a change form the way NASA does business, these three company's would also operate the spacecraft.

What resources can we use, from the Moon?

We would like to use lunar materials as much as possible, to avoid having to bring everything we need from Earth. This can represent a large cost reduction. It is called in-situ resource utilization. One thing we can certainly use is sunlight for power generation, with a strange 2 weeks on, 2 weeks off cycle. We will initially have to bring equipment and tools to get things going.

What does the moon have that we can use? Lots of rock (regolith), for local construction. We can think along the lines of in-situ resource utilization of local material for construction, and for a source of water, both for consumption and for rocket fuel (hydrogen and oxygen, or hydrogen peroxide) for the return trip. We can also return to Earth rare-earth elements. We might be able to manufacture solar cells directly from lunar material, which would address the local power problem.

Lunarcrete

If we build a structure on the moon, telescope on the backside or gambling casino, we will want to use local materials as much as possible. Everything else has to be hauled up from Earth. The term *lunarcrete* was coined in 1985 by Larry Beyer at the University of Pittsburgh. Not enough lunar material has been returned to Earth to check this in a lab. Several materials with similar property's have been developed under the broad title lunar regolith simulant. JSC has the largest batch of proof samples of real lunar material. Simulated material has been produced on the order of 30 metric tons. The real regolith varies in its properties, depending where it comes from on the lunar surface, such as Mare regions or lunar highlands.

Assuming we figure out the composition we will have to work with, a production facility can be constructed in situ. The material has to be brought in out of the vacuum for the process, since water will be involved (Ideally, water from local sources.)

Mixing the aggregate with cement, and introducing water, pre-cast concrete structures can be made. Same as on Earth, the resultant material will have limited tensile strength, and will require rebar or something similar. A local material, lunar glass, could also be used, which can also be formed from regolith. Another approach is to use lunar sulfur, which will bind the regolith. The mixture is heated to above the melting point of sulfur, which could be done outside, with solar mirrors. In tests at JSC, a mixture of 65% simulated regolith and 35 % sulfur worked well. The lower lunar latitudes reach temperatures above the melting point of sulfur during the lunar day, limiting the use of this approach. Another big deal with sulfur concrete is that it absorbs gamma rays, harmful to people and electronics. The resultant lunarcrete will need to be coated to make it airtight.

Observation spacecraft have found hints of large amounts of water ice at the poles, and in darkened craters. There may also be uranium, silicon, and metallic iron. We may find other useful stuff when we get our boots on the regolith.

Solar Cells

The necessary materials to produce solar cells are present in lunar soil. These include silicon, aluminum, and glass. We will have to build the initial facility on Earth (or, in orbit) to get the process started. The lunar material *anorthite* ($CaAl_2Si_2O_8$) can yield aluminum, calcium, silica glass, and some oxygen. There is also lunar iron, in conjunction with nickel, probably from meteorites. Some of the lunar polar regolith is known to contain hydrogen.

And lets not forget that the moon has little atmosphere, so we have access to all the vacuum we need, much better than what we can produce on Earth, and all for free.

Helium-3

Helium-3, is a non-radioactive isotope of helium. It is on the lunar surface, and is thought to have come from the solar wind. We're running low on helium on Earth – we used a lot for dirigibles, and it leaks through most materials, and rushes off into space. It would be valuable enough to send back to Earth. Helium-3 has the potential to enable a new concept energy source. In a fusion reactor, helium-3 releases a lot of energy, but nothing is radioactive. There are several possible reaction, like:

$$^2H + {}^3He \rightarrow He + {}^1p + 18.3 \text{ MeV}$$

This translates to an atom of hydrogen-2 (deutrium), and an atom of Helium-3 get together, and produce regular helium, a loose proton, and a fairly high level of energy. If we keep the proton trapped, there is no resultant radiation. This reaction actually generates enough power to sustain itself, and have power to spare. Another reaction is:

$$^3He + {}^3He \rightarrow {}^4He + {}^2p + 12.86 \text{ MeV}.$$

In this reaction, two Helium-3's get together, and form a Helium-4 (which is non-radioactive), an energetic proton, and power to spare.

We get a net energy gain from either reaction, and a spare energetic protons. This photon interacts with the confining field, and produces electricity as a by-product. Practically, the temperatures need to cause He3 to fuse are very high, and the whole process has to be contained, probably by magnetic fields. But, there are no "waste" products, and no dangerous radiation.

The Helium supply in the U. S. is under the control of the U. S. Department of Energy. Most of the Helium comes from the decay of Tritium in Nuclear warheads. There is a shortage, and it is critical for neutron radiation detectors, and for some medical diagnostics. The shortage is currently made up by the irradiation of lithium in a reactor. It is present in Earth's atmosphere, a little over 1 part per million. Helium-3 boils at around 3.2 degrees absolute.

One of the tasks of the Indian Lunar mission is mapping the lunar surface to Helium-3 concentrations. The Chinese Lunar Exploration Program is also very interested in lunar helium-3, and has made it a priority, as have the Russians.

Fusion plants on Earth, using Helium-3 and magnetic confinement could offer cost reductions. They would be cheaper to build and operate and certainly safer than current fission plants.

Lunar infrastructure

Communications with the Earth facing side of the moon is

relatively straightforward. On the backside, we need a communications relay satellite in a lunar polar orbit. The communications delay of about ½ a second is annoying, but can be tolerated.

The lunar surface has been mapped extensively, both before the Apollo missions to select suitable landing sites, and more recently by the Lunar Reconnaissance Orbiter. That NASA mission has been on the job since 2009, and it has accomplished a 3-D map of the lunar surface, with a resolution of 100 meters of more than 98% of the surface, missing only the polar regions. Apollo landing sites were captured in 0.5 meter resolution.

One piece of infrastructure the moon is missing is the equivalent of our GPS system on Earth, for precise surface position determination, and navigation. This can easily be solved. These satellites could also serve as communications relays for surface activities. We could in the short term use a ground-based LORAN type system for position determination.

There is an existing grid system of latitude and longitude for the lunar surface. What we will need is a timekeeping reference. We'll need a good 24 hour clock, since our bodies are attuned to that, and we can be in sync with Earth. Perhaps, GMT would be the best model. We also need a local time, although keep in mind a "day" and a "night" take about 2 weeks each. Still, it would be good to know when the sun would next rise, or set. That's fairly easy, and can be done in an ap.

A major resource of information on the Moon, Lava tubes, mining, and lunar surface activities, and all things lunar is the *Moon Miners' Manifesto*, written by Peter Koch, and hosted on the website of the National Space Society. Mr. Koch published ten issues per year from 1986 through 2012. His section on *Lunar Architecture and Construction* alone is 350 pages. See the Resources section in this book for a pointer.

Lunar Manufacturing

Manufacturing on the moon is not that far away, and has actually been done on a small scale for many years. With permanent manufacturing facilities in space, near to lunar or asteroid resources, we will be able to fabricate facilities from local material, and extract rocket fuel. All of this can replace what we now need very large rockets up from Earth's "gravity well." We can build the next generation stations and spacecraft in situ, in orbit. There are some major advantages to this, as fewer parts need to be lifted up to orbit. Spin-off company's, providing logistics services, are necessary. Space will be evolving as a frontier outpost. We have experience with those. But, space is a harsh environment, harsher even than the Klondike during the 1890's gold rush. Yet, the gold rush happened.

Viewed in an economic sense, manufacturing "space stuff" in space makes sense. There is a large initial investment, but the reduced costs, particularly of things that are intended to stay in space, will more than balance this out.

It is expected we will be able to separate out usable amounts of iron, aluminum, silicon, and oxygen from lunar and asteroid material. Extremely pure silicon wafers would be a valuable return-cargo. These could also be used in a subsequent process in-orbit to produce solar panels.

It is expensive to boost skilled workers up from Earth and keep them alive on the surface, but hopefully we can plan for some tele-operation from Earth, plus automated processes.

The rules change a bit in Space, and we will see yet another Industrial Revolution. One of the lessons learned in the first couple of Industrial Revolutions was that resources are rarely located where you need them. Manufacturing in Space is by all definitions a Paradigm Shift. NASA's Marshall Space Flight Center in Huntsville, Alabama hosts the National Center for Advanced Manufacturing. Many commercial players are in this topic, including Bigelow, Orbital Technologies, Excalibur-Almuz, Space

Island Group, and the Commercial Spaceflight Federation among others.

Power is free, once you set up the solar cells, and these could be built in space. If we need very high temperatures, we use focused sunlight. If we need very cold environments, we shield from direct sunlight. There is no ongoing cost for "hot" or "cold." The moon does have this consideration that we have warmth and sunlight for 2 weeks, and cold and darkness for two weeks.

Manufacturing bases can be isolated, so hazardous materials and processes could be done safely, tele-remotely. If the process goes badly, the pollutant is not on our home planet.

The company *Made-in-Space* is designing an advanced 3-D printer with robotic arms. This could fabricate and install structural members. The advantage of this approach is to have objects that won't fit on the launch vehicle, even with folding. The other advantage of in-space manufacturing is that the item won't have to survive launch vibration and acoustics. NASA is funding Made-in-Space, and its partners Northrop Grumman, and Oceaneering Space Systems.

Lunar base

The Shackleton Energy Company wants to exploit water ice of the moon. This is to lead to a network of refueling stations for ships with liquid propellant engines, using liquid water and liquid hydrogen. Bringing up the material from the surface of the moon is less costly in fuel, than from Earth. A study is needed on the relative economics of separating the hydrogen and oxygen on the lunar surface, and transporting the gases, or lifting the water to an orbiting hydrolysis station. The fuel and oxidizer would be available to government or other commercial entities (if you had their loyalty card.....3 points for every 10,000 gallons).

Shackleton is thinking of a fuel processing facility on the lunar surface, near the sources of water ice, with propellant depots in low Earth orbit. Hydrogen peroxide can also be manufactured, and is of value as a fuel.

Asteroid Amum 3554 is interesting to the Seattle-based company Planetary Resources. The M-class asteroid is about 1.6 km in size. It is estimated to contain a large amount of platinum, some $8 trillion ($10^{12}$) at current rates. Remember the Klondike gold rush? The asteroid also has massive amounts of iron, nickel, and cobalt. The project and company are well-financed by several entrepreneurial billionaires. We know the composition of the asteroid by spectral analysis. Near Earth asteroid 1986DA, about a mile wide, is estimated to contain 100,000 tons of platinum, and a mere 10,000 tons of gold. Can we gently and carefully nudge that to soft-land on the moon?

The Players

NASA is trying to engage and partner with commercial companies for Lunar exploration. The answer to the question. "What's in it for me?" might be, "You get to keep all the diamonds." The NASA cooperation will also focus on other nations as well. To the commercial world, the Moon is sometimes seen as the "8th continent." The technology of getting there and back has matured, and the race to exploit its resources is ongoing. Interest in the lunar surface is transitioning from governments to corporations.

This section discusses the companies who are active in developing approaches for lunar resources mining, and manufacturing In the haste to get this technology up and working, there are issues to be resolve, such as who owns what, who can exploit what, and who can be blamed. Similar to the Law of the Sea, there is a United Nations Law of Space, that most Nations have signed up to. It essentially says, you can't claim a celestial body. It's a bit vague on whether you can mine it for profit. Individuals and Companies are subject to the jurisdiction of their home country. Will this lead to small "safe-haven" country's luring big investment by hosting space mining companies to operate under their venue? Think, off-shore banking, Flag of Convenience. Small countries, with large and wealthy backers don't necessarily abide with International norms. Might we need an International Space Manufacturing Organization, modeled on the International Maritime Organization? If so, it's getting late to get it organized. It's about to become another "gold rush." There is, for example, no legal definition of where space begins.

Some of the existing organizations that address parts of the issue include the Inter-Agency Space Debris Coordination Committee; the Fourth Committee of the United Nations General Assembly, the Special Political and Decolonization committee; the United Nations Committee on the Peaceful Uses of Outer Space; and the United Nations Office for Space Affairs. None of these entities specifically address the ownership and usage of materials in space. There is an analogy to the exploitation of deep undersea mining. This is addressed in the U.N. Conventions on the Law of the Sea. There is an International Seabed Authority that has jurisdiction outside of a Nation's 200 mile Exclusive Economic Zone. Inside the zone, activities are regulated by the country's laws. There is also an International Institute of Space Law.

Space Law

Space law has its root in airspace law, from 1919. Who gets to fly over whose airspace, and who is responsible when you crash into some one else's country? Space policy was defined by the 1957 International Geophysical Year which saw the kick off of artificial satellites, overflying lots of other countries. The official governing document is called the "Treaty on Principles Governing the Activities of States in the Exploration and Use of Outer Space, including the Moon and Other Celestial Bodies."

You can get a Master of Law degree in Air and Space Law at several universities in the U.S. and other countries. That and a technical masters would make you very hire-able.

In addition, there are a number of relevant NGO's including the Commercial Spaceflight Federation, the European, Middle-East, and Africa Satellite Operators Association, and the Satellite

Industry Organization.

As of this writing, there is no law regarding extracted natural resources from space. Sovereignty over any real property above the Earth's atmosphere is not defined.

"...the governmental policy toward the private or commercial space sector will have a significant impact on the business chances of those private space ventures." Secure World Foundation.

Specifically for the Moon, there is the *Agreement Governing the Activities of States on the Moon and Other Celestial Bodies*, better known as the Moon Treaty. This has been signed by 17 countries. The UN has also put forth the Agreement Governing the Activities of States on the Moon and Other Celestial Bodies, 1979. This says that "no nation may claim sovereignty over any part of space. All countries should have equal rights to conduct research on the moon or other celestial bodies."

In addition, "All activities in space are required to be attached to a nation and any damages to other nations equipment or facilities caused by another party must be repaid in full to that nation." Thus, private company's must seek licenses from the relevant agency of their local governments to conduct space operations.

From the UN Committee on the Peaceful Uses of Outer Space, based on the International Law of the Sea, there have been derived five international treaties. These are:

- The 1968 Agreement on the Rescue of Astronauts, the Return of Astronauts and the Return of Objects Launched into Outer Space (the "Rescue Agreement").

- The 1972 Convention on International Liability for Damage Caused by Space Objects (the "Liability Convention").

- The 1975 Convention on Registration of Objects Launched into Outer Space (the "Registration Convention").

- The 1979 Agreement Governing the Activities of States on the Moon and Other Celestial Bodies (the "Moon Treaty").

- The 1967 Treaty on Principles Governing the Activities of States in the Exploration and Use of Outer Space, including the Moon and Other Celestial Bodies (the "Outer Space Treaty").

A total of 104 countries have agree to and signed the Outer Space Treaty.

Lunar Tourism

There have been spacefarers from over 40 countries, taken along on shared missions by the craft of the major spacefaring nations, China, Russia, and the U. S. There are new players in Space coming along, and commercial companies are competing in the New Frontier. The International Space Station is truly an International effort. But these were all professional Astronauts or Cosmonnauts. That was their job. They got paid for it.

At this writing, there have been seven "space tourists," who paid their own way, and five "spaceflight participants," who flew on the Shuttle, or to the ISS.

Several commercial company's are working toward the goal of space tourism, as the orbital and lunar infrastructure is being prepared for commercial purposes. The next step would be to use some of that infrastructure for paying tourists.

The Space Tourism Industry is just beginning. Like all new markets, it will evolve, become better and cheaper. It's expensive now, but a few have already done it. Options ranging from a quick trip above 100 km to earn the title "astronaut." to month-long vacations at a lunar resort. Space based casinos and athletic venues are on the drawing board.

NASA is not going to do this. They are in the science and technology business, and are a government agency, A cadre of entrepreneurs, space geeks, and crafty businessmen have better, less expensive options in the works. Stay tuned. Keep in touch. This is going to get exciting.

Keep your eye on commercial outfits engaged in Space Tourism. It is going to be huge. With many commercial companies building launch vehicles, and particularly re-usable launch vehicles, the cost of these adventures will come down. The moon is an obvious destination for a classy resort hotel.

The allure of space will generate the interest, which will be tempered by the high cost of access. Think what you can do at the Lunar hotel. You are nearly weightless. You can see the stars for weeks at time. You are looking down on Earth, how cool is that? In a lunar resort, you can fly. Really. With wings. At 1/6 gravity, the big lunar spare-air tank will allow you to experience human-powered, flapping-wing flight. I think chickens would really enjoy the opportunity to finally fly. And the penguins! Ok, even pigs could "fly."

Glossary

AESD – (NASA) Advanced Exploration Systems Division.
Apogee – furthest point in the orbit from the Earth.
Aphelion – furthest point to the Sun.
Apo-lune, or apo-selene - point of farthest approach to the lunar surface.
ARRM – asteroid retrieval robotic mission.
ASCE – American Society of Civil Engineers.
ASIN – Amazon Standard Inventory Number
Astrionics – electronics for space flight.
BEAM – Bigelow Expandable Activity Module – commercial inflatable space module.
BEO – beyond Earth orbit.
CATALYST - (Lunar) Cargo Transportation and Landing by Soft Touchdown
CBM – common berthing mechanism
CHM – cis-lunar habitation module
Cislunar – beyond Earth's atmosphere to just beyond the moon's orbit.
CLPS - (NASA's) Commercial Lunar Payload Services Program.
CM – crew module
CME – Coronal Mass Ejection, blast of energetic particles from the Sun.
CMP – co-manifested payload.
CNSA – China National Space Administration.
Conops – concept of operations.
COPOUS – (UN) Committee on the Peaceful Uses of Outer Space.
CPS – Cyrogenic Propulsion Stage.
CRTBP – Circular Restricted three-body Problem.
CSA – Canadian Space Agency, Agence Spatiale Canadienne
CSF – Cislunar Support Flight.
C&W – caution and warning.
Cygnus – Orbital-ATK automated cargo vehicle for ISS.
Cyrogenic – relating to very low temperatures.
DAM – damage avoidance maneuver.

DCM – docking cargo module.
Delta-V – change in velocity.
DoD – (U.S.) Department of Defense.
DoE – (U. S.) Department of Energy.
DRG – Distant Retrograde Orbit.
DRM – design reference mission.
DRO – distant retrograde orbit.
DSG – Deep Space Gateway
DSH – deep space habitat.
DSN – (NASA) Deep Space Network.
DST – Deep Space Transport
DTM – dynamic test model, for structural tests.
ECLSS – Environmental Control & Life Support system.
EDL – Entry, Descent, Landing.
EM-x Exploration Mission number-x.
Ephemeris – position information data set for orbiting bodies, 6 parameters plus time.
Epoch – a reference point in time for orbital elements.
EPS – electrical power system
ESA – European Space Agency
EUS – Exploration Upper Stage.
EVA – extra-vehicular activity.
Exosphere – low density "atmosphere", where the molecules don't act like a gas.
FAA – (U. S.) Federal Aviation Administration.
FMARS – Flashline Mars Arctic Research Station
GAM – Gateway Airlock Module
GLM – Gateway Logistics Module.
GMT – Greenwich Mean Time.
GNC – Guidance, Navigation, and Control.
GPO – (U.S.) Government Printing Office.
Gravity well – a conceptual model of the gravity field near a mass.
GSFC – NASA Goddard Space Flight Center, Greenbelt, MD.
Halo Orbit – three dimension orbit near the L1, L2, or L3 Lagrange points.
HEEO – highly eccentric Earth orbit.
HEOMD – (NASA) Human Exploration and Operations Mission

Directorate.
HITL – Human in the loop.
HOPE – Human Outer Planet Exploration (NASA)
HSIR – human systems integration requirements
IDSS – International Docking System Standard.
IGA - (ISS) InterGovernmental Agreement.
ILEWG - International Lunar Exploration Working Group.
ILOA - International Lunar Observatory Association.
ISP – specific impulse. Measure of efficiency of rocket engine. Units of seconds.
ISRO – Indian Space Research Organization
ISRU – in situ resource utilization
ISS – International Space Station
JAXA – Japan Aerospace Exploration Agency.
KW – kilowatt.
ISRU – in site resource utilization.
ISS – International Space Station
JAXA – Japanese space agency
JPL – Jet Propulsion Laboratory, Pasadena, CA.
JSC – Johnson Space Center, Houston, Texas.
KSC – NASA Kennedy Space Center, launch site, Florida.
L2 – second of 5 Lagrange points, a null in the gravity field in the restricted 3-body problem.
LADEE - Lunar Atmosphere and Dust Environment Explorer
LAS – launch abort system
Lbf – pounds, force.
LCT – Lunar Cargo Transportation.
LEO – Low Earth Orbit
LH2 – liquid hydrogen.
Libration point – null in the gravity field of the three body problem.
LOP-G – Lunar Orbital Platform – Gateway.
LORAN – radio navigation system, using fixed beacons.
LOS – Russian Lunar Orbital Station;m loss-of-signal.
LOX – liquid oxygen, boils at -297 F.
LRS – lunar regolith simulant.
LSAM – lunar surface access module

LSSPO – Lunar Surface Systems Project Office (NASA-JSC).
LST – landing (by) soft touchdown.
CATALYST – (lunar) Cargo Transportation and Landing by Soft Touchdown.
MARE, a dark basaltic plane, on the moon. From the Latin for sea.
MET – mission elapsed time.
Mev – million electron volts (unit of energy)
MMSEV – MultiMission Space Exploration Vehicle.
MOU – memorandum of understanding.
MPCV - Multi-Purpose Crew Vehicle.
MPLM – Multi-purpose Logistics Module.
m/s – meters per second.
Mt – metric ton, 1000 kg.
NAC – NASA Advisory Council.
Nadir – the point directly below.
NASA – (U.S.) National Aeronautics and Space Administration
NEO – near Earth object.
NextSTEP-2 – (NASA) Next Space Technologies of Exploration Partnerships.
NGO – non-profit, non-governmental organization.
NHV – net habitable volume.
NRHO – Near rectilinear halo orbit (around the L1 or L2 Earth-Moon libration point).
NSC – (U.S.) National Space council.
NSS – National Space Society.
NTIS – National Technical Information Service (www.ntis.gov).
NTRS – NASA Technical Reports Server, (ntrs.nasa.gov)
ORU – Orbital Replacement Unit.
OPSEK – (Russian) Orbital Piloted Assembly and Experiment Complex.
Perigee – closest point in the orbit from the Earth.
Perhelion – closest point to the Sun.
Perilune, or periselene – point of closest approach to the lunar surface.
PMA – Pressurized mating adapter.
PMCU – Power Management Control Unit.
PPB – power and propulsion bus.

PPE – power and propulsion element.
Precess – wobble, or have an angle that varies in a cyclic pattern.
PTCS – Passive thermal control system.
PVCU – Photo Voltaic Control Unit.
RCS – reaction control system.
RGA – rate gyro assembly
R&D – research & development.
Regolith – layer of loose material, covering rock; dirt.
RFI – request for information.
RMC – (NASA) Robotic Mining Competition.
ROSCOSMOS – Russian Space Agency.
RPOD – Rendezvous, Proximity Operations, Docking.
Selenodesy – science of studying and mapping the Moon's size, shape, and surface topography, as well as its gravitational and magnetic fields.
SEP – solar electric propulsion
SHFE – space human factors engineering.
SI – System International – the metric system.
Sidereal period – time for an object to make a full orbit.
Sol, local solar day – on the moon, about two Earth weeks.
SLS – (NASA) Space Launch System.
SMD – (NASA) Science Mission Directorate.
SPACE Act - Spurring Private Aerospace Competitiveness and Entrepreneurship.
Synodic period - time for an object in orbit to occupy the same point, in relation to 2 other objects.
TCS – thermal control system.
TLI – Trans-lunar injection.
TM – Technical Manual.
TPS – thermal protection system.
Trillion - 10^{12}
TRL – technology readiness level.
UDM – universal docking module.
ULA – United Launch Alliance, joint venture between Lockheed Martin and Boeing.
Ullage – residual fuel or oxidizer in a tank after engine burn is complete.

UN – (Earth) United Nations.
USAF – United States Air Force.
VTVL – vertical takeoff, vertical landing.
V&V – verification and validation.
WDV – water delivery vehicle.
XBASE - Expandable Bigelow Advanced Station Enhancement.
Zenith – the point directly above.

References

Aldrin, Buzz *No Dream Is Too High, Life Lessons From a Man Who Walked on the Moon*, National Geographic, 2016, ISBN-9781426216497.

Benaroya, Haym *Building Habitats on the Moon: Engineering Approaches to Lunar Settlements,* (Springer Praxis Books), 2018, ISBN-3319682423.

Berton, Pierre *The Klondike Fever: The Life And Death Of The Last Great Gold Rush*, 2015, ASIN-B06XGD1TCX.

Burke, J. D. "Development of a lunar infrastructure," Acta Astronautics, Vol. 17, Issue 7, July 1988. avail: ScienceDirect.

Buss, Jared S. *Willy Ley: Prophet of the Space Age*, 2017, University Press of Florida, ISBN-0813054435.

Crawford, Ian "Lunar Resources: A Review," in Publication, avail: https://arxiv.org/ftp/arxiv/papers/1410/1410.6865.pdf

Daglis, I. A. (Editor), 2001, *Space Storms and Space Weather Hazards*, Springer-Verlag New York, ISBN-1-4020-0031-6.

Eckhart, Peter *The Lunar Base Handbook,* 1999, 1st ed, McGraw-Hill Primis Custom Publishing, ASIN-B01A1MSBRK.

Edwards, Bradley C. and Westling, Eric A. *The Space Elevator: A Revolutionary Earth-to-Space Transportation System,* 2003, ISBN-097465171.

Fassbender, Melissa "Drug Development in space: how microgravity enables pharma," 2016, avail: http://www.outsourcing-pharma.com/Preclinical-Research/Eli-Lilly-drug-development-in-space

Fowler, Wallace *Moon Port: Transportation Node in Lunar Orbit: NASA's Effort to Support a Manned Lunar Colony*, 2015, ASIN – B014LRVBOQ.

Grey, Jerry *Space Manufacturing Facilities (Space Colonies)*, 1977, AIAA, ASIN-B000VODUG6.

Gurtuna, Ozgur *Fundamentals of Space Business and Economics*, 2013, SpringerBriefs in Space Development, ISBN-1461466954.

Harris, Phillip *Space Enterprise: Living and Working Offworld in the 21st Century,* 2009, ASIN-B00DZ0PDPO.

Haley, Andrew Gallagher, *Space Law And Government,* 2012, ISBN-1258267152.

Häuplik-Meusburger, Sandra Olga Bannova, Olga *Space Architecture Education for Engineers and Architects: Designing and Planning Beyond Earth* (Space and Society), 2016, Springer, ISBN-9783319192796.

Heiken, Grant, (ed) eta al, *Lunar Sourcebook: A User's Guide to the Moon,* 1991, Cambridge University Press, ISBN-0521334446

Heimreich, Robert L. "The undersea habitat as a space station analog evaluation of research and training potential" (SuDoc NAS 1.26:180342), 1985.

Holliday, J. S. *The World Rushed In: The California Gold Rush Experience,* 2002, ISBN-080613464X.

Hudgins, Edward L *Space: The Free-Market Frontier*, 2003, Cato Institute, ASIN-B005HITTR0.

Krukin, Jeff, *NewSpace Nation: America's Emerging Entrepreneurial Space Industry, 2002,* 2nd Edition, ASIN-B00719IM12.

Lewis, John S. *Asteroid Mining 101: Wealth for the New Space Economy*, 2014, Deep Space Industries, ASIN-B01LP8JMNQ.

Matloff, Greg, Bang, C. *Harvesting Space for a Greener Earth*, 2014, ISBN-1461494257.

Matson, John "Is MOON's Sci-Fi Vision of Lunar Helium 3 Mining Based in Reality?" Scientific American – News Blog, Jun 12, 2009.

Mendell, Wendell W. *Lunar bases and Space Activities of the 21st century,* 1985, Lunar and Planetary Institute, ISBN 0-942862-02-3.

Merrow, Mark *A Lunar Space Station: NASA's Study to Design a Lunar Space Station in Support of a Manned Moon Base,* 2015, alc Books, ASIN-B014LQ177S.

NASA, *NASA Space Technology Report: Lunar and Planetary Bases, Habitats, and Colonies, Special Bibliography Including Mars Settlements, Materials, Life Support, Logistics, Robotic Systems,* ASIN-B00CLX44E2.

NASA Systems Engineering Handbook, NASA SP-2007-6105.

NASA, *The Sun, the Earth, and Near-Earth Space: A Guide to the Sun-Earth System - Comprehensive Information on the Effects of Space Weather on Human Life, Climate, Spacecraft,* 2013, ISBN-9780160838071. Avail: https://bookstore.gpo.gov/, stock number 033-000-01328-1.

NASA, *Lunar Colonization: Energy and Power*, 2012, ISBN 1288290675.

O'Neill, Gerald K. AIAA, *Space-Based Manufacturing from Nonterrestrial Materials,* 1977, Progress in Astronautics and Aeronautics, ISBN-10-0915928213.

O'Neill, Gerard K., O'Leary, Brian, "Lunar Resources and their Utilization", Space-Based Manufacturing from Nonterrestrial Materials, Progress in Astronautics and Aeronautics, pp. 97-123. htttps://doi.org/10.2514/5.9781600865312.0097.0123.

Radley, Charles; Pearson, Jerome *The Lunar Elevator: Bringing the Riches of the Moon Down to Earth*, Springer. 2018, ISBN-3319664867.

Sacksteder, Kurt R.; Sanders, Gerald B. "In-situ resource utilization for lunar and mars exploration,". AIAA Aerospace Sciences Meeting and Exhibit 2007, ISBN-1-62410-012-3.

Santarius, John "Lunar He3 and Fusion Power," 2004, avail: http://fti.neep.wisc.edu/presentations/jfs_ieee0904.pdf

Schrunk, David; Sharpe, Burton *The Moon: Resources, Future Development and Settlement* (Springer Praxis Books), 2007, ISBN-0387360557.

Seedhouse, Erik *Lunar Outpost: The Challenges of Establishing Human Settlement on the Moon,* 2008, Springer, ISBN-0387097465.

Schwab, Klaus *The Fourth Industrial Revolution*, 2017, Crown Business, ASIN-B01JEMROIU.

Slyuta, E. N.; Abdrakhimov, A. M.; Galimov, E. M. (March 12–16, 2007).The Estimation of Helium-3 Probable Reserves in Lunar Regolith PDF). 38th Lunar and Planetary Science Conference. p.2175. Avail:
avail: https://www.lpi.usra.edu/meetings/lpsc2007/pdf/2175.pdf

UN, International Space Law: United Nations Instruments, 2018, ISBN-9211013844.

U.S. Government, *U.S. Interpretation of International Space*

Policies Regarding Commercial Resource Acquisitions - Evolving Space Laws and Treaties, Legalizing Commercial Space Mining on the Moon and Asteroids, 2018, ISBN-1981076476.

Vance, Ashlee *Elon Musk: Tesla, SpaceX, and the Quest for a Fantastic Future*, 2015, Ecco, ISBN-0062301233.

Mark Williams (August 23, 2007)."Mining the Moon: Lab experiments suggest that future fusion reactors could use helium-3 gathered from the moon," avail: https://www.technologyreview.com/s/408558/mining-the-moon/

Zubrin, Robert *Entering Space: Creating a Spacefaring Civilization*, 2000, ISBN-10-1585420360.

Resources

- NASAspaceflight.com
- http://www.nasa.gov/offices/education/centers/kennedy/technology/nasarmc.html
- Space Studies Institute, www.ssi.org
- http://www.spaceislandgroup.com/manufacturing.html
- Partnerships to Advance the Business of Space, Sep 3, 2014 by Subcommittee on Science and Space of the Committee on Commerce, Science, and Transportation United States Senate.
- http://www.commercialspaceflight.org/
- Secure World Foundation, Handbook for New Actors in Space, avail: https://swfound.org/handbook/
- https://humanizing.tech/the-coming-gold-rush-of-space-manufacturing-601d9c2dd8b6
- International Institute of Space Law, www.iislweb.org
- Space Act, 2015 – avail: https://www.congress.gov/bill/114th-congress/house-bill/2262

- https://www.space.com/28189-moon-mining-economic-feasibility.html
- https://space.nss.org/lunar-resources-unlocking-the-space-frontier/
- https://cedb.asce.org/CEDBsearch/record.jsp?dockey=0057886
- https://www.thealternativedaily.com/valuable-resources-moon/
- https://arc.aiaa.org/doi/abs/10.2514/5.9781600865312.0097.0123
- https://www.theceomagazine.com/business/innovation-technology/mining-the-moon-the-space-start-ups-looking-for-lunar-resources/
- https://www.space.com/28189-moon-mining-economic-feasibility.htm
- https://www.theceomagazine.com/business/innovation-technology/mining-the-moon-the-space-start-ups-looking-for-lunar-resources/
- https://www.masten.aero/
- http://nssdc.gsfc.nasa.gov/planetary/lunar/surveyor.html
- Civil Engineering Database, Lunar Resources, avail: https://cedb.asce.org/CEDBsearch/record.jsp?Dockey=0057886.
- NASA-AMES, In-Situ Resource Utilization, https://www.nasa.gov/centers/ames/research/technology-onepagers/in-situ_resource_Utiliza14.html
- http://www.moonexpress.com/
- https://space.nss.org/lunar-base-and-settlement-library/
- https://space.nss.org/moon-miners-manifesto/
- wikipedia, various

You might also be interested in some of these.

Stakem, Patrick H. *16-bit Microprocessors, History and Architecture*, 2013 PRRB Publishing, ISBN-1520210922.

Stakem, Patrick H. *4- and 8-bit Microprocessors, Architecture and History*, 2013, PRRB Publishing, ISBN-152021572X,

Stakem, Patrick H. *Apollo's Computers*, 2014, PRRB Publishing, ISBN-1520215800.

Stakem, Patrick H. *The Architecture and Applications of the ARM Microprocessors*, 2013, PRRB Publishing, ISBN-1520215843.

Stakem, Patrick H. *Earth Rovers: for Exploration and Environmental Monitoring*, 2014, PRRB Publishing, ISBN-152021586X.

Stakem, Patrick H. *Embedded Computer Systems, Volume 1, Introduction and Architecture*, 2013, PRRB Publishing, ISBN-1520215959.

Stakem, Patrick H. *The History of Spacecraft Computers from the V-2 to the Space Station*, 2013, PRRB Publishing, ISBN-1520216181.

Stakem, Patrick H. *Floating Point Computation*, 2013, PRRB Publishing, ISBN-152021619X.

Stakem, Patrick H. *Architecture of Massively Parallel Microprocessor Systems*, 2011, PRRB Publishing, ISBN-1520250061.

Stakem, Patrick H. *Multicore Computer Architecture*, 2014, PRRB

Publishing, ISBN-1520241372.

Stakem, Patrick H. *Personal Robots*, 2014, PRRB Publishing, ISBN-1520216254.

Stakem, Patrick H. *RISC Microprocessors, History and Overview,* 2013, PRRB Publishing, ISBN-1520216289.

Stakem, Patrick H. *Robots and Telerobots in Space Application*s, 2011, PRRB Publishing, ISBN-1520210361.

Stakem, Patrick H. *The Saturn Rocket and the Pegasus Missions, 1965,* 2013, PRRB Publishing, ISBN-1520209916.

Stakem, Patrick H. *Visiting the NASA Centers, and Locations of Historic Rockets & Spacecraft,* 2017, PRRB Publishing, ISBN-1549651205.

Stakem, Patrick H. *Microprocessors in Space*, 2011, PRRB Publishing, ISBN-1520216343.

Stakem, Patrick H. Computer *Virtualization and the Cloud*, 2013, PRRB Publishing, ISBN-152021636X.

Stakem, Patrick H. *What's the Worst That Could Happen? Bad Assumptions, Ignorance, Failures and Screw-ups in Engineering Projects, 2014,* PRRB Publishing, ISBN-1520207166.

Stakem, Patrick H. *Computer Architecture & Programming of the Intel x86 Family, 2013,* PRRB Publishing, ISBN-1520263724.

Stakem, Patrick H. *The Hardware and Software Architecture of the Transputer*, 2011,PRRB Publishing, ISBN-152020681X.

Stakem, Patrick H. *Mainframes, Computing on Big Iron*, 2015, PRRB Publishing, ISBN- 1520216459.

Stakem, Patrick H. *Spacecraft Control Centers*, 2015, PRRB Publishing, ISBN-1520200617.

Stakem, Patrick H. *Embedded in Space,* 2015, PRRB Publishing, ISBN-1520215916.

Stakem, Patrick H. *A Practitioner's Guide to RISC Microprocessor Architecture*, Wiley-Interscience, 1996, ISBN-0471130184.

Stakem, Patrick H. *Cubesat Engineeering*, PRRB Publishing, 2017, ISBN-1520754019.

Stakem, Patrick H. *Cubesat Operations*, PRRB Publishing, 2017, ISBN-152076717X.

Stakem, Patrick H. *Interplanetary Cubesats*, PRRB Publishing, 2017, ISBN-1520766173 .

Stakem, Patrick H. Cubesat Constellations, Clusters, and Swarms, Stakem, PRRB Publishing, 2017, ISBN-1520767544.

Stakem, Patrick H. *Graphics Processing Units, an overview*, 2017, PRRB Publishing, ISBN-1520879695.

Stakem, Patrick H. *Intel Embedded and the Arduino-101, 2017,* PRRB Publishing, ISBN-1520879296.

Stakem, Patrick H. *Orbital Debris, the problem and the mitigation*, 2018, PRRB Publishing, ISBN-*1980466483*.

Stakem, Patrick H. *Manufacturing in Space*, 2018, PRRB Publishing, ISBN-1977076041.

Stakem, Patrick H. *NASA's Ships and Planes*, 2018, PRRB Publishing, ISBN-1977076823.

Stakem, Patrick H. *Space Tourism*, 2018, PRRB Publishing, ISBN-

1977073506.

Stakem, Patrick H. *STEM – Data Storage and Communications*, 2018, PRRB Publishing, ISBN-1977073115.

Stakem, Patrick H. *In-Space Robotic Repair and Servicing*, 2018, PRRB Publishing, ISBN-1980478236.

Stakem, Patrick H. *Introducing Weather in the pre-K to 12 Curricula, A Resource Guide for Educators*, 2017, PRRB Publishing, ISBN-1980638241.

Stakem, Patrick H. *Introducing Astronomy in the pre-K to 12 Curricula, A Resource Guide for Educators*, 2017, PRRB Publishing, ISBN-198104065X.
Also available in a Brazilian Portuguese edition, ISBN-1983106127.

Stakem, Patrick H. *Deep Space Gateways, the Moon and Beyond*, 2017, PRRB Publishing, ISBN-1973465701.

Stakem, Patrick H. *Exploration of the Gas Giants, Space Missions to Jupiter, Saturn, Uranus, and Neptune*, PRRB Publishing, 2018, ISBN-9781717814500.

Stakem, Patrick H. *Crewed Spacecraft*, 2017, PRRB Publishing, ISBN-1549992406.

Stakem, Patrick H. *Rocketplanes to Space*, 2017, PRRB Publishing, ISBN-1549992589.

Stakem, Patrick H. *Crewed Space Stations*, 2017, PRRB Publishing, ISBN-1549992228.

Stakem, Patrick H. *Enviro-bots for STEM: Using Robotics in the pre-K to 12 Curricula, A Resource Guide for Educators*, 2017, PRRB Publishing, ISBN-1549656619.

Stakem, Patrick H. *STEM-Sat, Using Cubesats in the pre-K to 12 Curricula, A Resource Guide for Educators*, 2017, ISBN-1549656376.

Stakem, Patrick H. *Lunar Orbital Platform-Gateway*, 2018, PRRB Publishing, ISBN-1980498628.

Stakem, Patrick H. *Embedded GPU's*, 2018, PRRB Publishing, ISBN- 1980476497.

Stakem, Patrick H. *Mobile Cloud Robotics*, 2018, PRRB Publishing, ISBN- 1980488088.

Stakem, Patrick H. *Extreme Environment Embedded Systems,* 2017, PRRB Publishing, ISBN-1520215967.

Stakem, Patrick H. *What's the Worst, Volume-2*, 2018, ISBN-1981005579.

Stakem, Patrick H., *Spaceports*, 2018, ISBN-1981022287.

Stakem, Patrick H., *Space Launch Vehicles*, 2018, ISBN-1983071773.

Stakem, Patrick H. *Mars*, 2018, ISBN-1983116902.

Stakem, Patrick H. *X-86, 40th Anniversary ed*, 2018, ISBN-1983189405.

Stakem, Patrick H. *Lunar Orbital Platform-Gateway*, 2018, PRRB Publishing, ISBN-1980498628.

Stakem, Patrick H. *Space Weather*, 2018, ISBN-1723904023.

Stakem, Patrick H. *STEM-Engineering Process*, 2017, ISBN-1983196517.

Stakem, Patrick H. *Space Telescopes,* 2018, PRRB Publishing, ISBN-1728728568.

Stakem, Patrick H. *Exoplanets*, 2018, PRRB Publishing, ISBN-9781731385055.

Stakem, Patrick H. *Planetary Defense*, 2018, PRRB Publishing, ISBN-9781731001207.

Patrick H. Stakem *Exploration of the Asteroid Belt*, 2018, PRRB Publishing, ISBN-1731049846.

Patrick H. Stakem *Terraforming*, 2018, PRRB Publishing, ISBN-1790308100.

Patrick H. Stakem, *Martian Railroad,* 2019, PRRB Publishing, ISBN-1794488243.

Patrick H. Stakem, *Exoplanets,* 2019, PRRB Publishing, ISBN-1731385056.

Patrick H. Stakem, *Exploiting the Moon,* 2019, PRRB Publishing, ISBN-1091057850.

Patrick H. Stakem, *RISC-V, an Open Source Solution for Space Flight Computers,* 2019, PRRB Publishing, ISBN-1796434388.

Patrick H. Stakem, *Arm in Space*, 2019, PRRB Publishing, ISBN-1099789133.

Patrick H. Stakem, *The Search for Extraterrestial Life,* 2019, PRRB Publishing, ISBN-1072072181.

www.ingramcontent.com/pod-product-compliance
Lightning Source LLC
Chambersburg PA
CBHW030735180526
45157CB00008BA/3184